I0469853

Ten
Together

Written and illustrated by ORNA

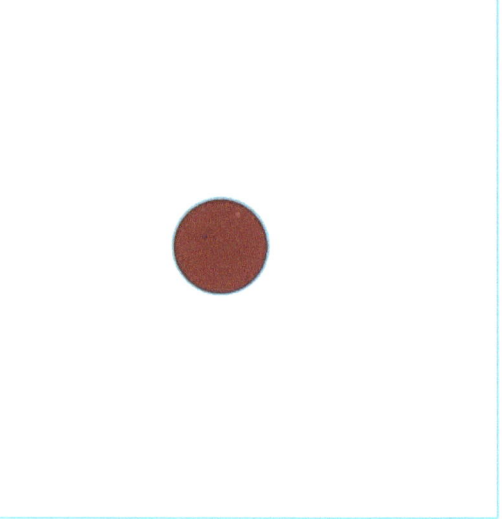

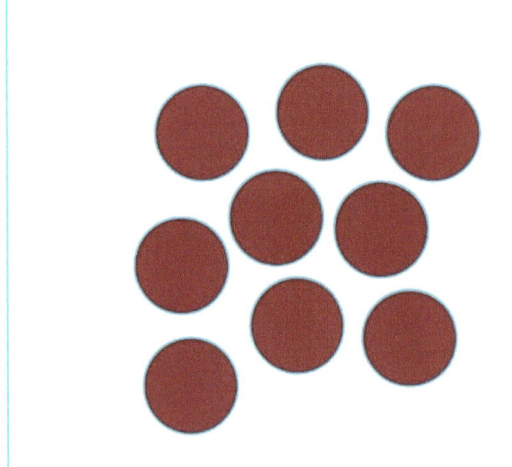

One

plus

Nine

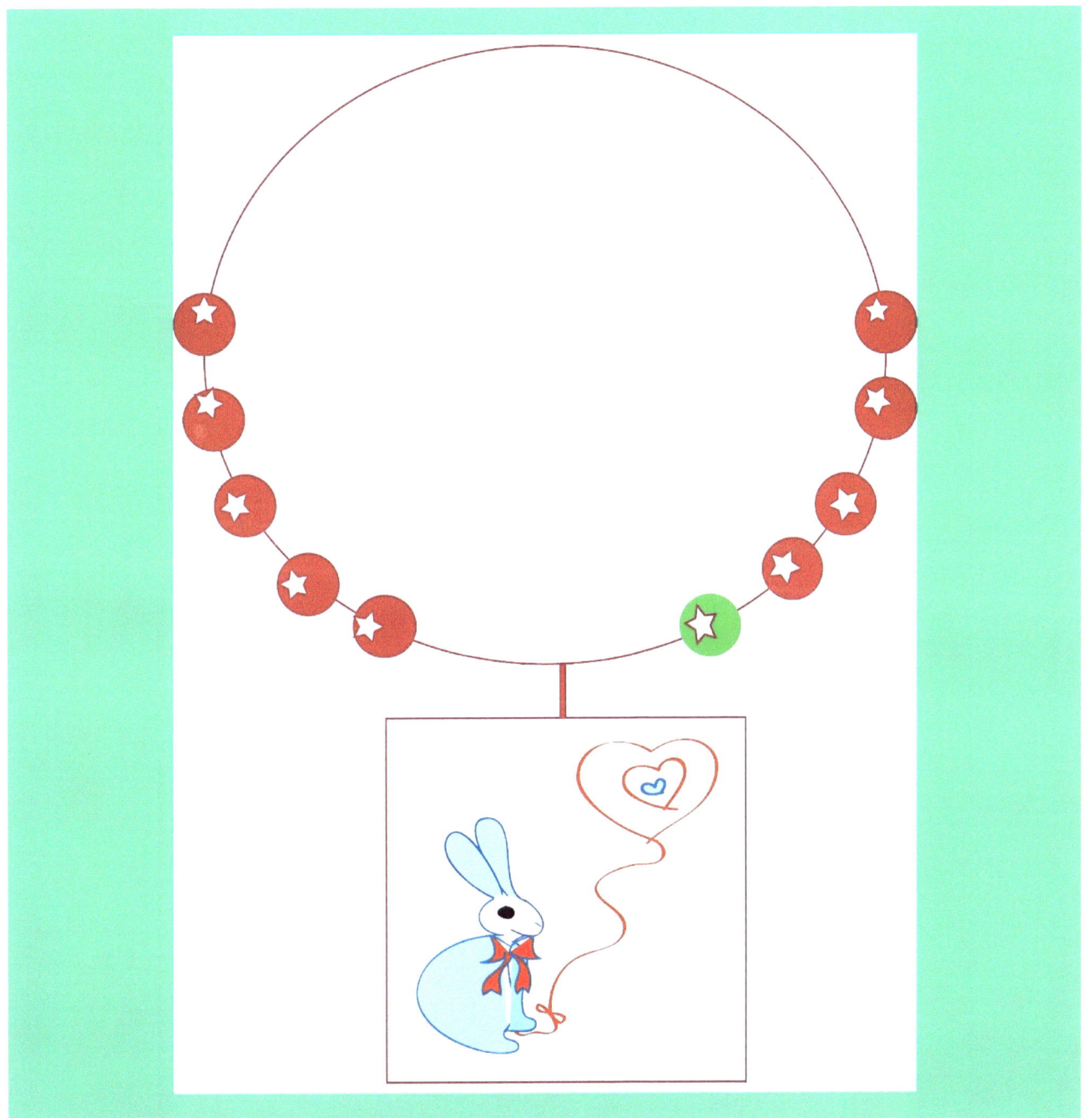

Two

plus

Eight

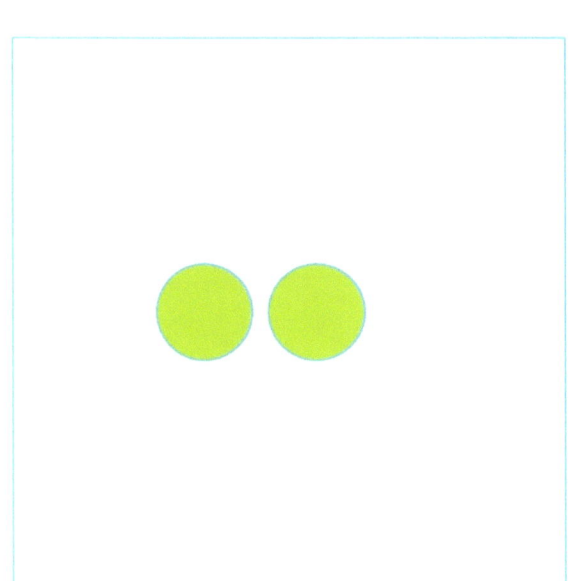

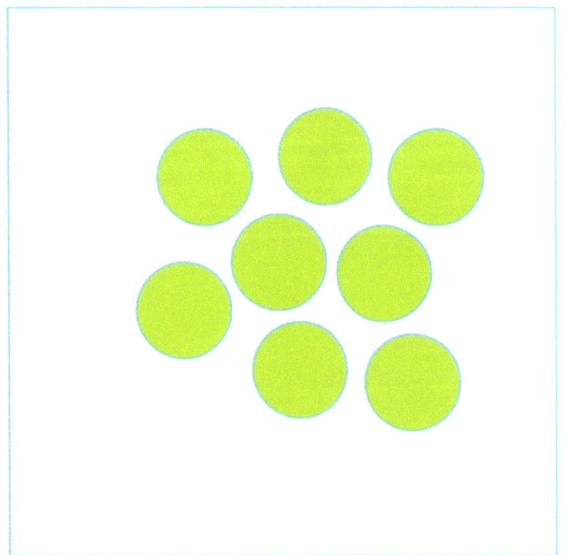

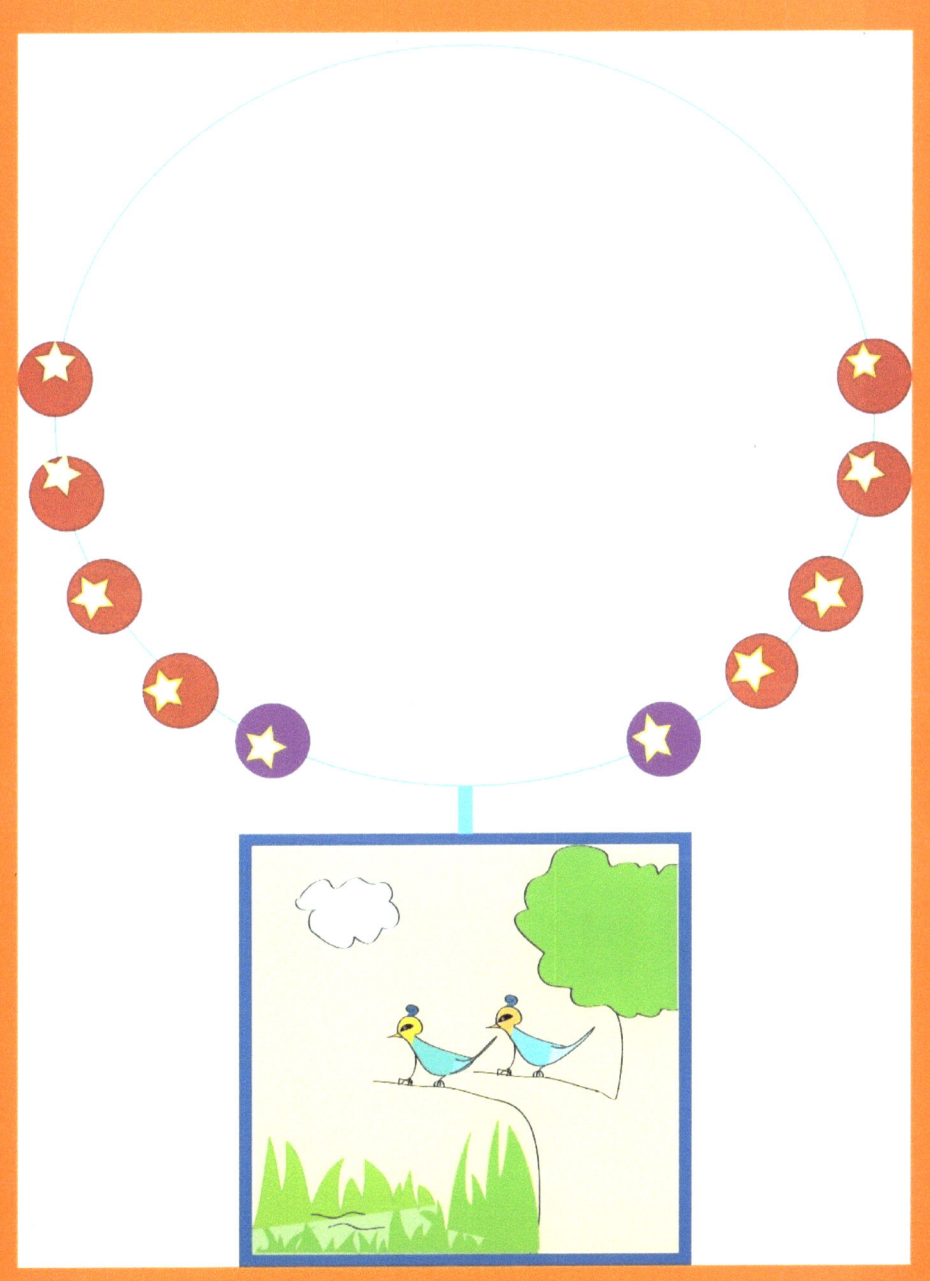

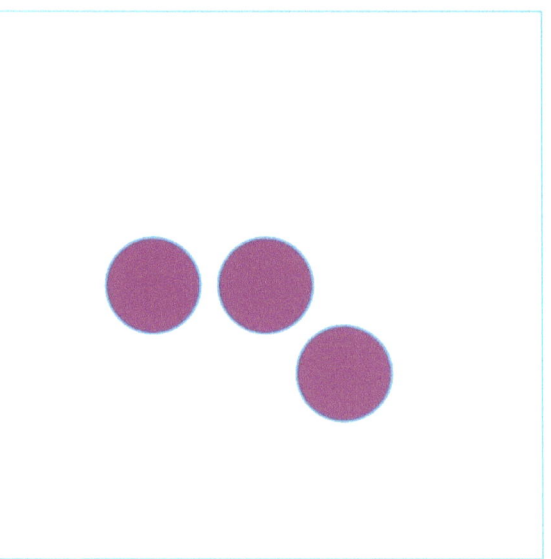

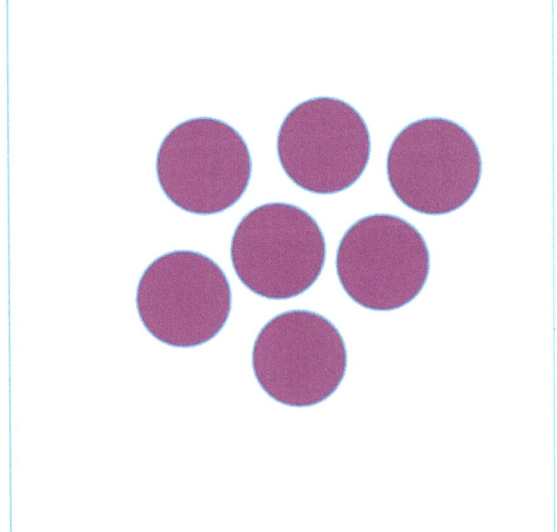

Three

plus

Seven

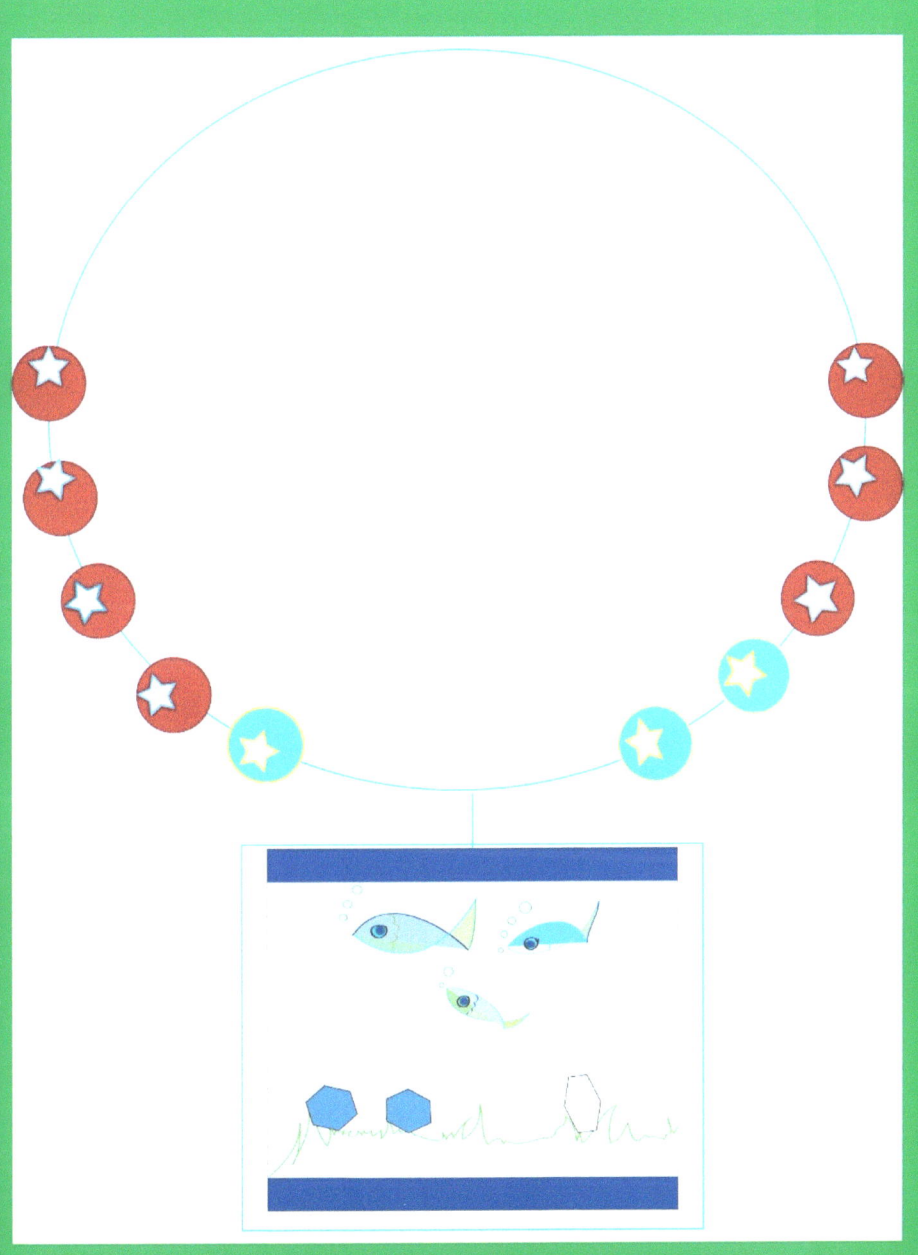

Four

plus

Six

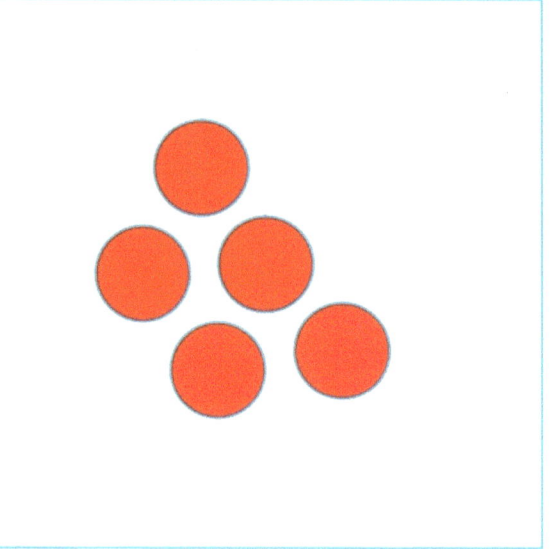

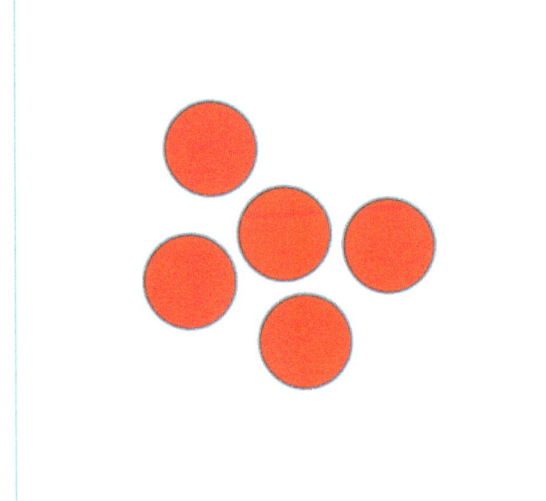

Five

plus

Five

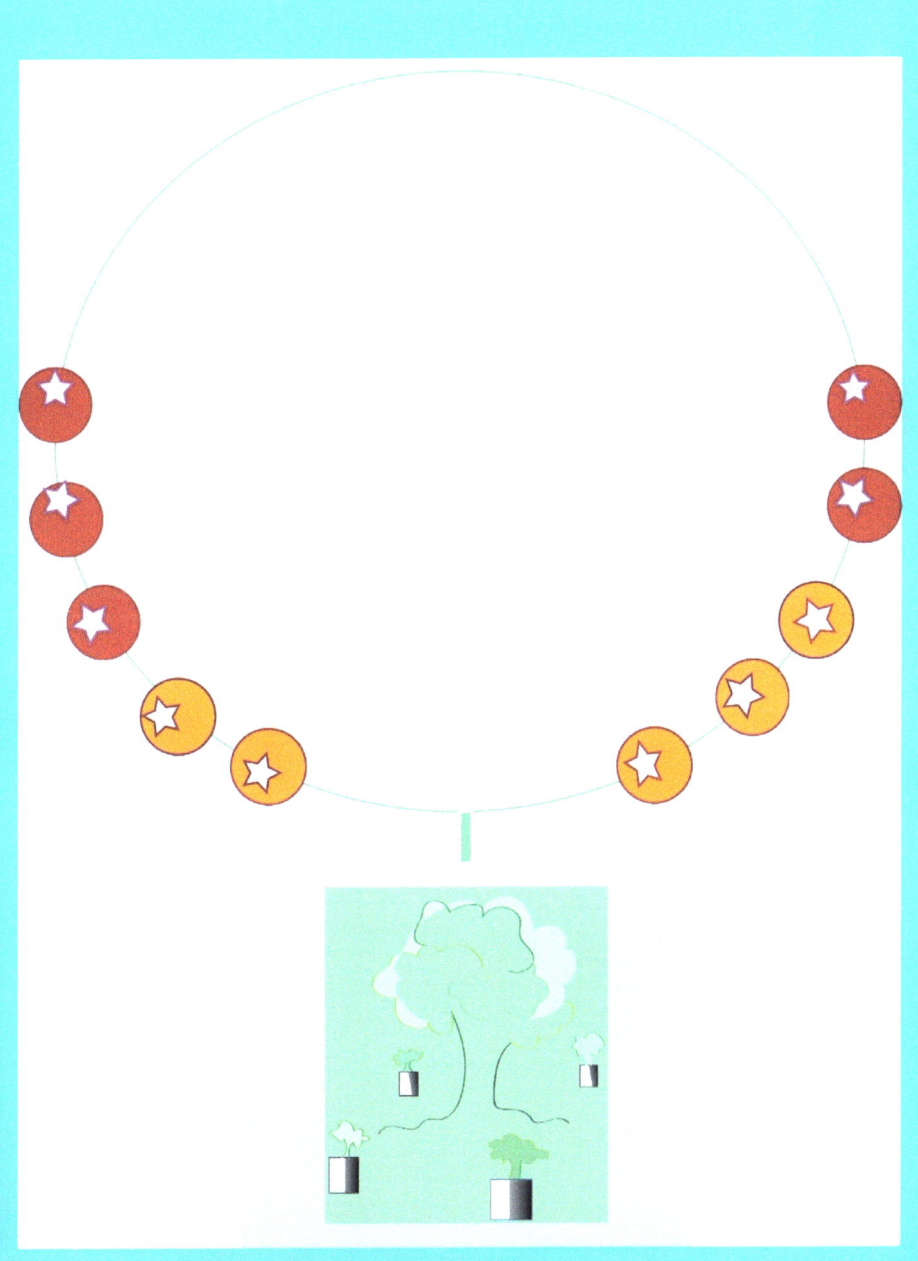

Six

plus

Four

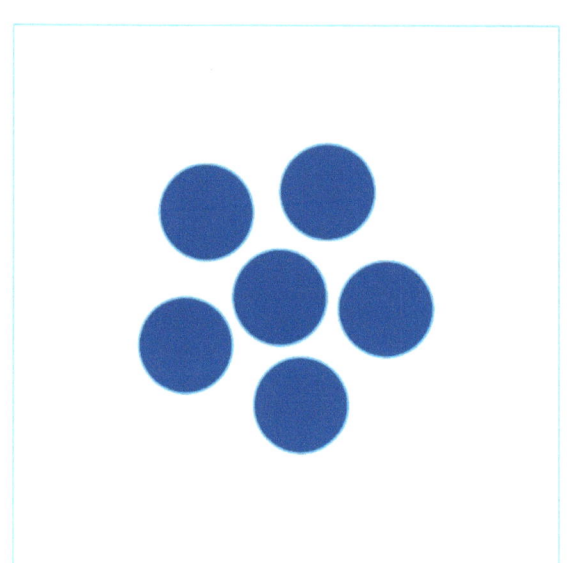

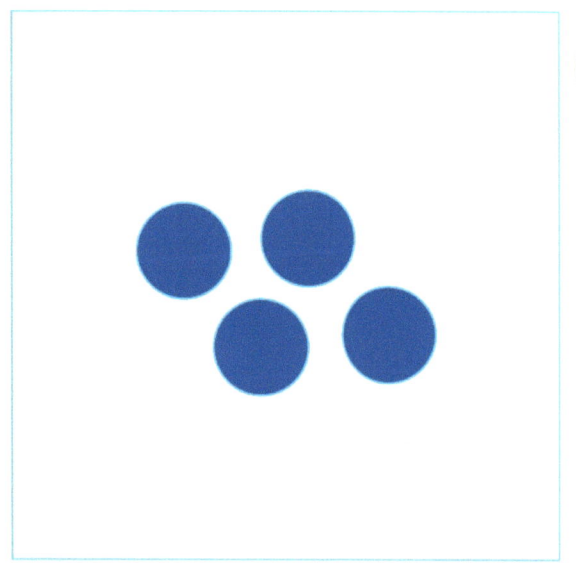

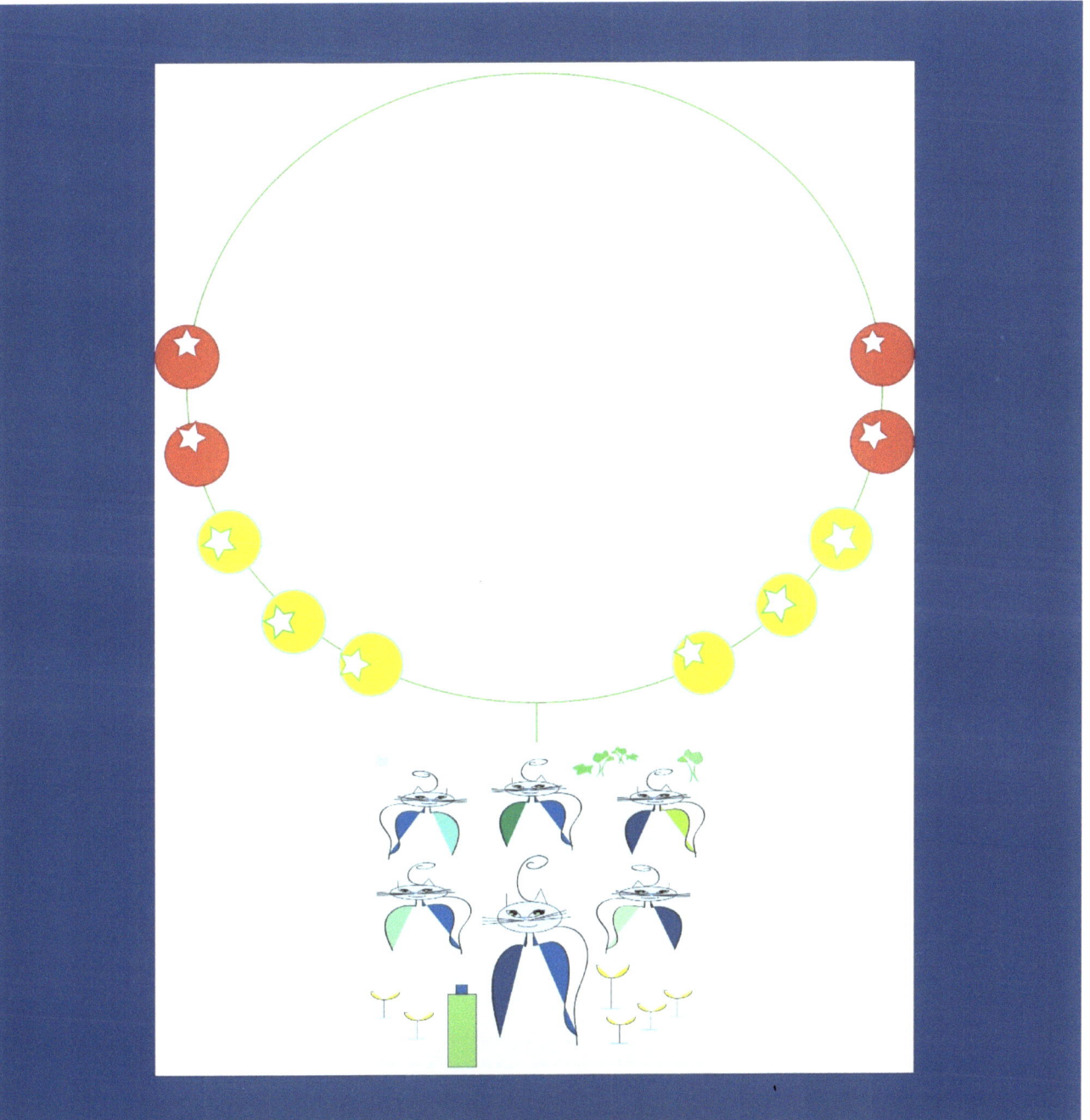

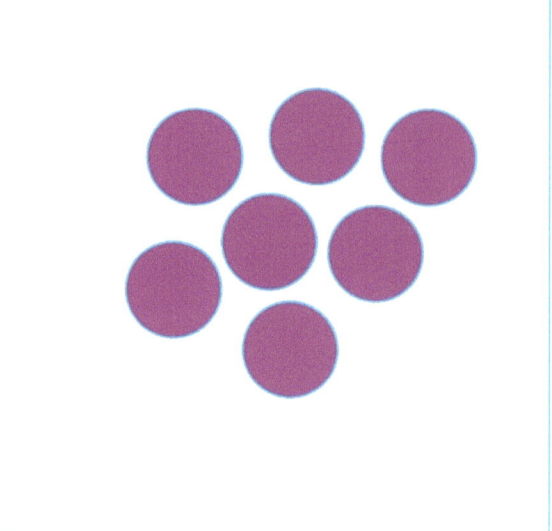

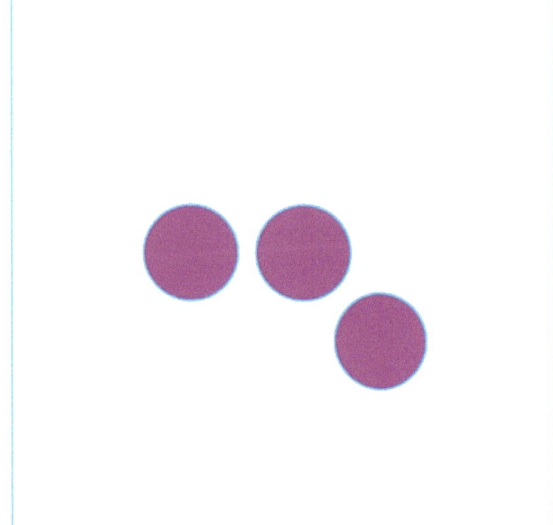

Seven

plus

Three

Eight

plus

Two

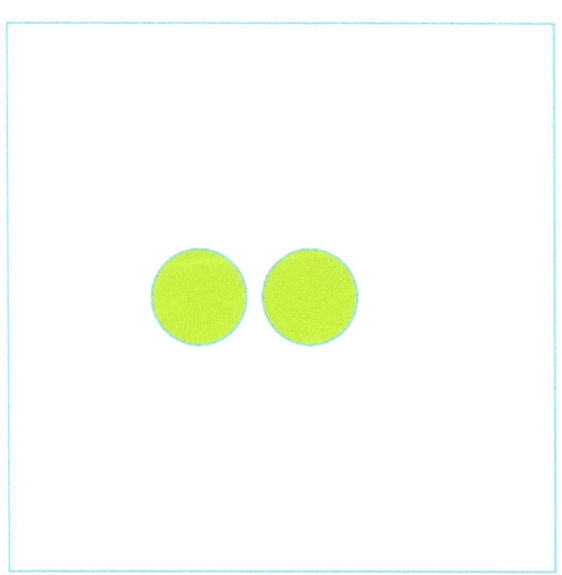

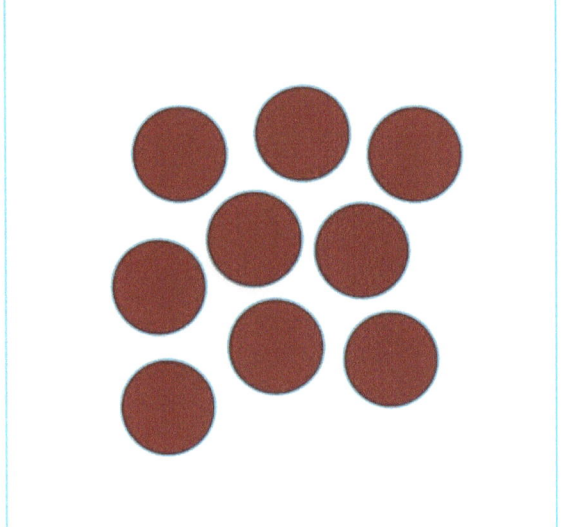

Nine

plus

One

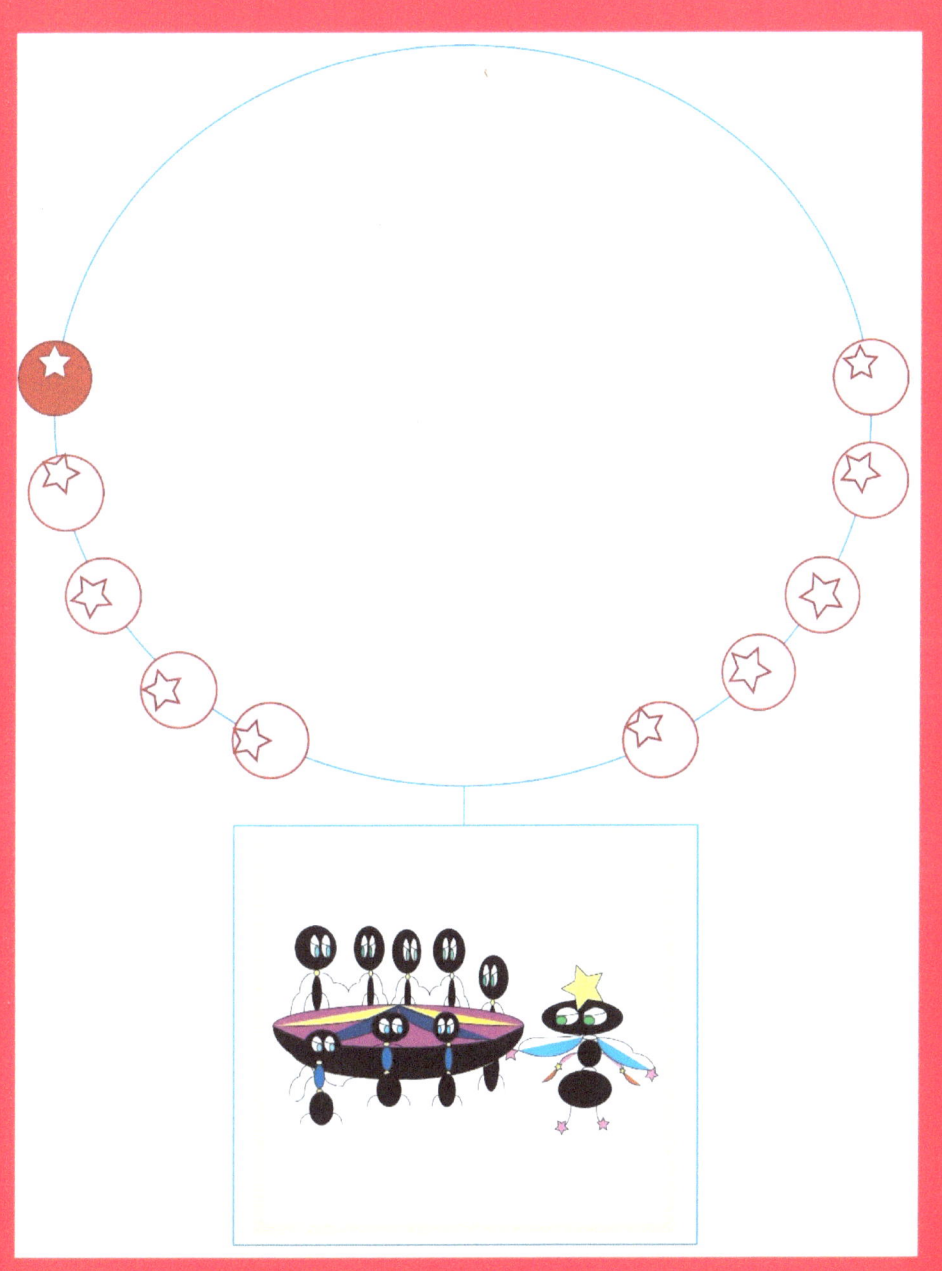

Ten

plus

Zero

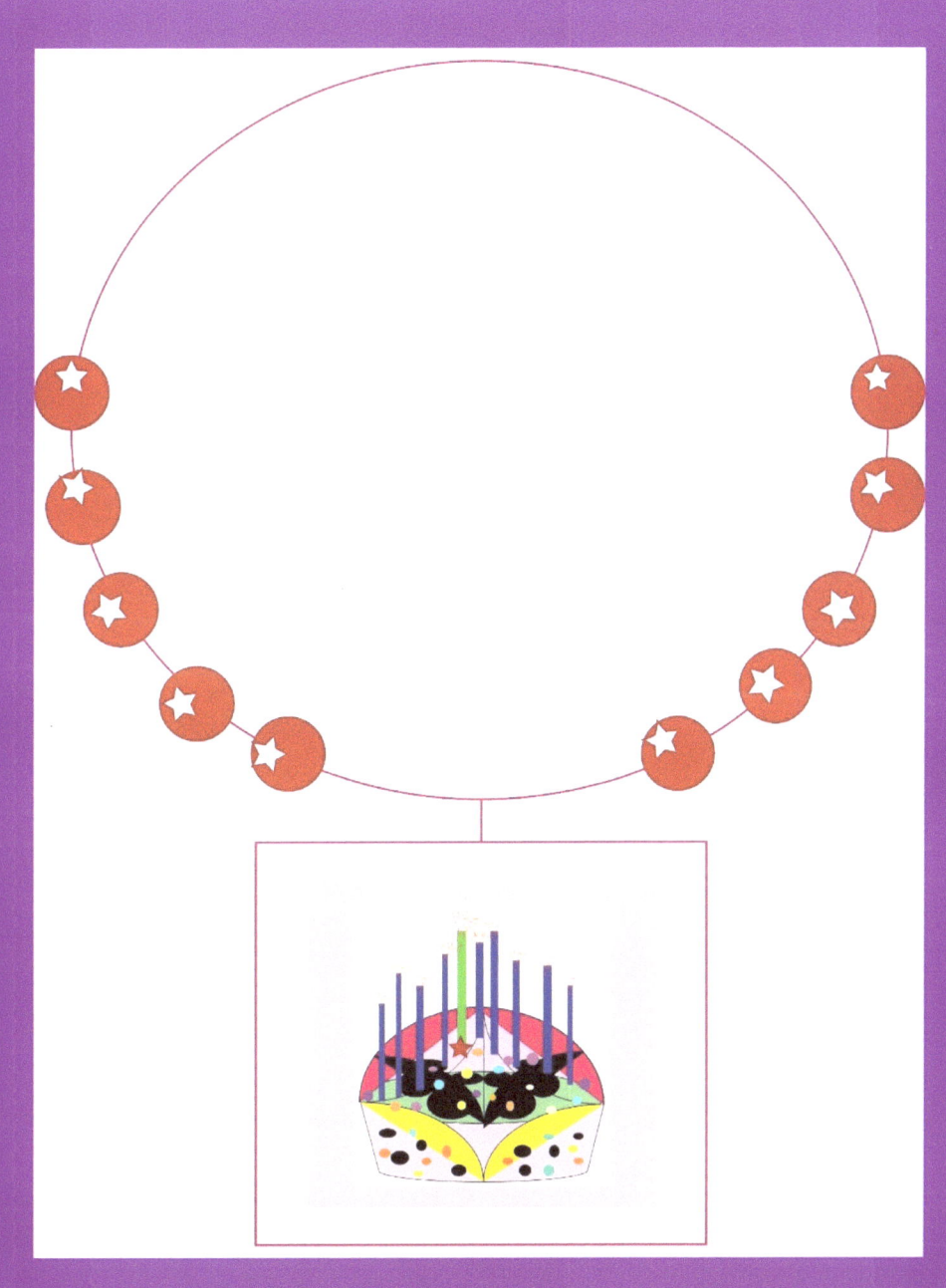

Here
you may
attach a photo,
a drawing, etc.
for
the number
ten.

Colored by:

Ten
Together

Copyright © 2015 ORNA

www.ingramcontent.com/pod-product-compliance
Lightning Source LLC
Chambersburg PA
CBHW050432180526
45159CB00006B/2507